GRAVITY AND ROTATION

AUDREY E. RANDLES

Copyright © 2012 Audrey Elizabeth Randles
All rights reserved.

Cover Image: 'Blue Marble - Image of the Earth from Apollo 17'
Image credit: NASA

Introduction

The Theory of Matrix series of books offers the exiting developments in cosmological theory. We combine elements of psychology, cosmology, and astrophysics to discover secrets hidden deep in the Universe.

NASA's Gravity Probe B confirms two Einstein Space-Time theories 'Imagine the Earth as if it were immersed in honey. 'As the planet rotates, the honey around it would swirl, and it's the same with space and time,' said Francis Everitt, GP-B principal investigator at Stanford University. 'GP-B confirmed two of the most profound predictions of Einstein's universe, having far-reaching implications across astrophysics research. Likewise, the decades of technological innovation behind the mission will have a lasting legacy on Earth and in space.' NASA

Our current understanding of gravity did not lead us to antigravity exploration. To resolve these issues, we analyse the Matrix of the Earth. The Theory of Matrix provides us with an explanation of the fundamental processes behind Newton's Law of Universal Gravitation.

'Gravity and Rotation' is the 4th book of the series. In this book, we summarise the general part of the Theory of Matrix related to the specific aspects of gravity and gravitational acceleration, antigravity and rotation of high energy massive radiating systems and the associated rotation of cosmic background radiation.

We analyse the specific geometry of the Earth and the influence of the Matrix forces.

In 1920 Ernst Gehrcke made a point, 'besides other things, it remained unanswered, where this energy of this gravitational field is coming from.' We hope to solve the riddle in this book.

Stay well, and enjoy your reading.

Yours sincerely,

Audrey Elizabeth Randles

JULY 28, 2020

Image 1: 'Blue Marble - Image of the Earth from Apollo 17'
Image credit: NASA

Contents

Introduction
Acknowledgement
Chapter   1 Multimodal Space-Time
Chapter   2 Space-Time Structure of the Earth
Chapter   3 Energy Structure of the Earth
Chapter   4 Space-Time and Info-Energy Imbalance
Chapter   5 Matrix Forces
Chapter   6 Gravity of the Earth
Chapter   7 Antigravity of the Earth
Chapter   8 The Equivalence Principle
Chapter   9 Transmission of Gravity and Antigravity
Chapter 10 Rotation of the Earth
Afterword
Content Use Policy

Acknowledgement

We would like to express our gratitude to the National Aeronautics and Space Administration (NASA), NASA's Goddard Space Flight Center (GSFC), Solar Dynamics Observatory, NASA's Jet Propulsion Laboratory (JPL), the California Institute of Technology (Caltech), the University of California, Irvine (UC Irvine), Arizona State University, and the University of California at Los Angeles (UCLA) for the excellent images and exciting descriptions, published by NASA's Jet Propulsion Laboratory (JPL) and used in this book to illustrate the spectacular beauty of our dynamic world, its complex structures, and success of the space explorations addressing fundamental questions of the Earth physics and human psychology concerning our place in the Universe.

The views and opinions of the author, expressed in this book, do not necessarily state or reflect those of NASA's Goddard Space Flight Center, Solar Dynamics Observatory, NASA's Jet Propulsion Laboratory, the California Institute of Technology, the University of California, Irvine, Arizona State University, the University of California at Los Angeles or National Aeronautics and Space Administration.

Multimodal Space-Time

We introduce a new understanding of the world as the multimodal world with reference to the multimodal, multidimensional Space-Time and Info-Energy structure of the Universe and the existing objects and systems.

Every existing single object and every system of the objects, including visually 'empty' spaces, subatomic particles and holes, stars and planets, galaxies and star clusters, systems holding Black Holes, and the thermodynamic Universe, are the multimodal objects and systems. The multimodal objects and systems carry the complex properties of multiple Space-Time modalities. Characteristics of the objects and systems are various in different modalities of Space-Time.

The 1-dimensional, 2-dimensional, and 3-dimensional elements of an object or a system are integrated into the object or the system's multimodal Space-Time and Info-Energy structure. We call the multimodal Space-Time and Info-Energy structure of an object or a system - the Matrix for short. The Matrix depicts the total mass and energy of an object or a system in space and time, including its past, present, and future.

Prof. Hermann Weyl described the Matrix of Light - the Light Cone of the Special Relativity, as follows,

'According to the relativity principle, however, the bisection of past-future is of a different kind when it is seen from world point 0, and it corresponds to the one that in three-dimensional space is caused by a complete circular cone (it is sketched in the vertical projection in the figure; the curved line is the world line of my body, that is of course bisected through 0 into two parts, the past part and the future part of my life).' [Einstein Albert, Lenard Philipp, Weyl Hermann, Gehrcke Ernst, The Bad Nauheim Debate. The Discussion concerning the theory of relativity at the Meeting of Natural Scientists. (1920)]

The Matrix of the Earth depicts the multimodal Space-Time, Info-Energy, and Mass-Energy structure of the Earth and, accordingly, the structure of the Matrix involves a coexistence of different Space-Time modalities.

The meaning of time applies to the Matrix settings as the past, future, and current time of the system, existing 'here' 'now', in the period of time now occurring. The Matrix reflects our momentary perception of time as a point 'now', associated with the period of time now occurring, and the current human understanding of two separate imperceptible periods associated with the past and future.

A '0' Space-Time point at the centre of the Matrix is a point 'here' and 'now'. The point 'here' and 'now' as a '0' point was mentioned in Special Relativity in association with the mathematical model of the Light Cone. 'Let us call any world-point 0 as a space-time-null-point.' [Minkowski Hermann, 'Space and Time' (1920)] 'A world-point is a 'here-now'. [Weyl Hermann, 'The Discussion concerning the theory of relativity at the Meeting of Natural Scientists' (1920)]

The '0' Space-Time point may be represented as a coexistence of the '0' time-point and '0' space-point, which are complementary, building the time system and the space system in the opposite direction. A '0' time-point - the point 'now', reflects a time component of the '0' Space-Time point.

A 'o' space-point - the point 'here', reflects a space component of the 'o' Space-Time point.

The 'o' Space-Time point is the point of the Matrix symmetry if the Matrix is balanced in its Space-Time and energy.

Energy draws space and time. The Matrix of the Earth reflects the current human understanding of energy as the forms of energy acting in our dynamic world as radiation, mass, and kinetic energy, and the latent form of potential energy.

The system's potential Info-Energy is arranged by the 2-dimensional representations of background radiation and light into the system's 2-dimensional Info-Energy grid forming the system's time and potential space. Any object, which is smaller than the wavelength supporting the Matrix grid of the object, does not exist in Space-Time.

According to Albert Einstein'...the ponderable masses will be the determining factor in producing the field, or, according to the fundamental result of the special theory of relativity, the energy density...' [Albert Einstein, A Brief Outline of the Development of the Theory of Relativity (1921)]

The Matrix is not the field but the structure. The ponderable masses are the determining factors in producing the energy density, information density, and multimodal Space-Time and Info-Energy structures of the existing objects and systems. The Matrixes for the objects and systems of limited mass and energy are limited in space and time.

The Matrix, being a property of the object or the system, does not exist without the object or the system. The associated object or the system is a source of the total Space-

Time and Info-Energy of its Matrix and, nevertheless, the Matrix is a self-organising system.

Matrixes are generated as an outcome of the binary operation of two Matrixes or by any of the following: reflection, self-duplication, binary fission, self-multiplication, and incorporation while retaining anisotropy.

Image 2: 'Earthrise over the lunar horizon'
Image Credit: NASA/Goddard/Arizona State University

Space-Time Structure of the Earth

Two types of the Matrixes have been identified - the Time-Rising Matrix (TRM) and the Space-Rising Matrix (SRM).

The TRMs are the property of the radiating objects and systems, including 'black body' radiation spectrum. The 2-dimensional grid of the TRM has a form of the Riemannian Manifold with the negative curvature (Figure 1).

According to the theory of General Relativity, energy curves Space-Time. Please see Einstein Albert, 'Relativity: The Special and General Theory' (1916). We support this idea and provide descriptions of the main mechanisms as precise as possible for a person functioning, as we are, in the dynamic 3-dimensional world.

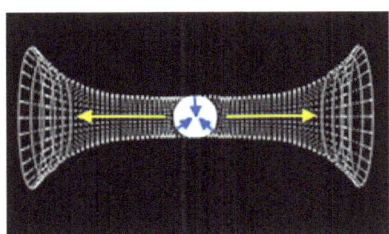

Figure 1: Time-Rising Matrix

The Matrix grid, forming the system's multimodal Space-Time, is typically a rectangular, regular repeated, with the equal distances, 2-dimensional infinitely thin filament. It is the 2-dimensional arrangement of the object or the system's latent Info-Energy. The 2-dimensional grid, shaping the Earth's multimodal Matrix, limits the volume of the system at the Matrix centre.

Strictly speaking, the reproduction of the Matrix grid on the paper is not exact though the forms of the Matrixes are correct. It is related to the difficulties to reproduce the 2-dimensional structure on the paper. Please notice that we rely entirely on descriptions.

The 'o' time-point, representing the time component of the 'o' Space-Time point at the centre of the Matrix, reflects the human perception of time as a point 'now' and provides us with an opportunity to perceive the volumes of the objects and systems existing in our dynamic world.

Space-Time is represented in the Earth multimodal Space-Time and Info-Energy structure as the Spaces of Time. The meaning of time applies to the Matrix settings as the past, future, and current time of the system existing 'now', in the space of the period of time now occurring.

The Space of the Current Time is associated with the period of time now occurring. The Space of the Current Time at the centre of the Matrix is formed by the 2-dimensional grid and connected with the Spaces of the Progressive and Regressive Time.

It is enclosed with the 2-dimensional Info-Energy grid at the Matrix centre. It equals the volume of the Earth in the period of time now occurring. The volume of the Earth is bounded by the 2-dimensional grid at a 'o' time-point at the centre of the TRM.

Our understanding of the objects characteristics, motion, acceleration, and forces, acting in our dynamic world, applies to the Space of the Current Time.

In the General Theory of Relativity, Albert Einstein described gravity as a consequence of the curvature of Space-Time caused by the uneven distribution of mass. In astrophysics, an event horizon is a boundary beyond which events cannot affect an observer. In our dynamic world, we operate in the Space of the Current Time.

As we sum up the ideas behind these statements, we can introduce the boundary of the curvature of the system's Space-Time as an event horizon that surrounds the total volume of the system that equals the system's Space of the Current Time. As such, the external boundary of the curvature of the Earth Space-Time affected by the Earth gravity is the external boundary of the Earth Space of the Current Time.

In our descriptions, we refer the boundary of the curvature of the system's Space of the Current Time affected by gravity as 'the surface of the system'. Besides, the specific toroid-globe geometry and the associated uneven distribution of mass are the cause of the particular geometry of the Earth in our dynamic world.

The 2-dimensional grid connects the Space of the Current Time with the Spaces of the Progressive and Regressive Time.

The Matrix Space of the Progressive Time is the space of the Matrix cone associated with the future of the system (Figures 1-4 - the Matrix left cone). The Space of the Regressive Time is the space of the Matrix cone associated with the past of the system (Figures 1-4 - the Matrix right cone). The Spaces of the Progressive and Regressive Time look like a hologram. They are immobile, unchanging, inaccessible, appeared empty, and connected with the undifferentiated Continuum. Please see my book 'Space and Time' for details.

Following the relative difference in the human mode of perception of time and space, we cannot perceive any movement and changes of the 1-dimensional Space-Time of the Continuum, as a black background on the images of the Matrixes (Figures 1-4), and the Matrix cones.

There is no priority of the front cone holding the future events nor the back cone holding the past events in the description of the Light Cone of the Special Relativity. There is neither priority of the Progressive nor Regressive Time in the Matrix of an object or a system. The Spaces of the Progressive and Regressive Time are symmetrical via the Space-Time axis and 'o' Space-Time point at the Matrix centre if the Matrix is balanced. Every Matrix demonstrates a tendency to obtain and retain a balance.

Under the Theory of Special Relativity, the space axis of the Light Cone builds a perpendicular to the time axis. In compliance with our investigations of the Matrix and according to the Theory of Matrix, the Light Cone is an example of the Matrix for a flash of light. The Space-Time axis (other terms: Matrix axis, space axis, time axis) is the only axis of the Matrix. The Matrix is symmetrical via the Matrix axis if the Matrix is balanced in its Space-Time, Mass-Energy, and Info-Energy.

The Matrix Space-Time Arrow, directed along the Space-Time axis, demonstrates a coexistence of the tendency of time and the contra-directed tendency of space, along with the tendency of the past and the contra-directed tendency of the future.

The time component of the Space-Time Arrow might be represented as the Arrow of the Progressive Time and the Arrow of the Regressive Time. The Arrows of Time are associated with the Spaces of the Progressive and Regressive Time.

The Arrows of Time are contra-directed, symmetrical, and balanced if the Matrix is balanced. They indicate the direction of the time and time-associated energy 'flow' in the Matrix Spaces of Time.

The Arrows of Time are drawn as the yellow arrows on the reproduction of the Matrixes (Figures 1-4), while the blue arrows represent the direction of the Space of the Current Time in accordance with the human perception of the objects and systems as displaying three spatial dimensions x, y, and z, or a combination of three directions. In case of a high energy massive radiating system, such as the Earth, the system's Arrows of Space start at the external boundary of the curvature of the system's Space-Time affected by gravity.

In the TRM of the Earth, the Space of the Progressive Time is a deviation from the Progressive Time with the characteristics of the time and time-associated latent energy 'flow' and the Arrow of the Progressive Time directed from a '0' time-point to the future of the Earth.

The Space of the Regressive Time is a deviation from the Regressive Time with the characteristics of the time and time-associated latent energy 'contra-flow' and the Arrow of the Regressive Time directed from the '0' time-point to the past of the object.

The Matrix Arrows of Time may be reversed that would indicate the reverse of the direction of the time and energy flow in the Matrix Spaces of Time and the reverse of Space-Time of the multimodal system, such as a planet, the Sun or a radiating galaxy. Space-Time reversibility is determined by the Space-Time and Info-Energy resources of the system. If the system reaches the limits of its Space-Time and Info-Energy imbalance, it will be reversed by the Matrix vector-forces. The speed of light in a vacuum defines the exact proportion of space to time as the upper limit of our dynamic world, and Planck length, Planck time, and Planck energy define the exact lower limits for an object or a system's existence.

The direction of the Arrows of Time allows us to detect possible changes in the Space-Time direction and reverse of the time and energy flow in the objects and systems of the objects.

Image 3: 'Moon - False Colour Mosaic' Image credit: NASA/JPL

Energy Structure of the Earth

Energy is a quality of information, and information is a quality of energy. The information represents a form of energy 'in formation', in dynamics. Information is always associated, processed, stored, or transmitted along with energy or matter. Energy exists in a form set by information, and information has a form fixed by energy. The density of information is proportional to the density of energy associated with the information. We use the term 'Info-Energy' (forms of energy) to describe the energy and information, which are complementary and cannot be separated.

Mass-Energy is 'the mass of a body regarded as energy, according to the laws of relativity' (Oxford Dictionary). We, respecting the laws of relativity, use the term 'Mass-Energy' as interchangeable with the term 'Info-Energy' of an object or a system.

Energy and information are the fundamental characteristics of the Universe and the existing objects and systems. Info-Energy is to be defined in relation to a frame of reference.

Nothing is coming from nothing and going to nothing. Energy and information cannot be lost. Energy exists in the latent, or potential, and actual forms. The information takes different forms and exists in the latent, or potential, and actual forms along with energy.

Space and mass are associated with the actuality. You can directly perceive the volume and weight of the book you keep in your hand.

Time and time-related Info-Energy are latent to us. We cannot directly perceive time-related potential energy and associated information. Time and tide wait for no man. Some objects and systems, events and their changes lie in the past. For example, we cannot right now taste the dinner that we had the other night. On the other hand, we cannot, at this point, read a book we will first glance at tomorrow. Both the dinner and the book are latent to us - we cannot touch or see them directly at the moment.

Time and time-related Info-Energy are associated with potentiality as a capacity to perform work.

When the object or the system's capacity to perform work is being realized - energy is associated with space. Similarly, water, filling up an empty container, takes shape. Energy, existing in space, takes perceivable forms creating sensible information. We sense different forms of energy as masses, kinetic energy, and other energy representations actually acting in our dynamic world in the period of time now occurring.

The total Info-Energy of an object (or a system) is the total energy and associated information, which the existing object possesses in space and time, including its past, present, and future.

The object's Space of the Current Time is filled with the actual Info-Energy. We can directly perceive and measure the objects and systems' volumes and the forms of actual energy, such as masses, kinetic energy, and other forms of energy acting in the volumes of objects and systems in our dynamic world in the period of time now occurring. The complete actual information is coded and fixed by the object's actual energy and represented in forms of energy and matter in the volume of the object. The density of information is proportional to the density of energy and matter associated with the information.

The properties of the Earth are reflected in the Matrix characteristics. The multimodal structure of the Earth is limited in its volume, mass, energy, and the time of existence.

The volume of the Earth is filled with the actual Info-Energy. The total actual Info-Energy of the Earth is an integrated form of energy and associated information acting in the volume of the Earth in the period of time now occurring. The total actual energy of the Earth is coded and fixed by the complete actual information and represented in mass, kinetic energy, and other energy representations acting in the volume of the Earth in the period of time now occurring.

The multimodal structure of the Earth is formed by the 2-dimensional grid. The Matrix grid depicts the Space-Time and Info-Energy structures of the Earth.

Albert Einstein writes: 'It was found that inertia is not a fundamental property of matter, nor, indeed, an irreducible magnitude, but a property of energy.' [Einstein Albert, A Brief Outline of the Development of the Theory of Relativity (1921)].

We mentioned above that the 2-dimensional grid is a regular repeated rectangular arrangement of the info-energetic net-structure. It is infinitely thin filament forming the Matrix Spaces of Time and containing the Earth latent and potential energy and associated information. We can compare the 2-dimensional grid with a human genome that is the complete set of genetic information for a newborn child.

The 2-dimensional grid, forming the Matrix cones, contains the time-associated 2-dimensional potential energy of the system. It looks like an immobile and unchanging holographic image. We cannot have a clear perception of the 1-dimensional and 2-dimensional time structures of the multimodal objects and systems. The time-associated 2-dimensional Info-Energy of the 2-dimensional grid forms the time-limits of the Earth existence. It forms the Spaces of Time. Energy, building time, is latent Info-Energy associated with the past, present, and future of the planet.

The space-associated latent 2-dimensional Info-Energy of the 2-dimensional grid composes the fundamental 2-dimensional Space-Time of the Matrix. The Matrix grid limits the object's volume at the centre of the Matrix and shapes the object's Spaces of Time.

The space-associated latent 2-dimensional Info-Energy of the 2-dimensional grid forms the potential space of the period of time now occurring. The Earth potential space underlies the volume of the Earth within its multimodal Space-Time and Info-Energy structure.

The potential space of the Current Time forms the 2-dimensional 'container' for an object or a system's volume filled with actual Info-Energy. The 2-dimensional grid, forming the Matrix Space of the Current Time is easily perceivable by the observer located at the '0' time-point at the centre of the Matrix and the observer located on the outside.

The total Info-Energy of the Matrix grid is the total latent, or potential, energy and associated latent information, that the Earth possesses in space and time, including its past, present, and future.

The TRM of the Earth has a form of the Riemannian Manifold with the negative curvature, and Riemannian geometry applies to the Matrix investigation. The curvature of the Matrix grid depends on the Space-Time, Info-Energy, and Mass-Energy properties of the Earth. The curvature of the grid does not depend on how the surface is embedded in 3-dimensional or higher-dimensional spaces. The curvature of the grid does not depend on how the 3-dimensional space is embedded at the centre of the Matrix.

Black pixels, reading zero, corresponding to the potential information, are fixed by the potential energy of the 2-dimensional grid. They are associated with the timing mechanisms and sweep rates, and the address of a pixel corresponds to its Space-Time coordinates.

We suppose that the blocks of information, which are involved in the actuality or located relatively close to the point 'now' in time, are situated closer to the centre of the TRM then other latent information. Hypothetically, if the blocks of information are associated with the Space of the Current Time, then they might have the 3-dimensional structure.

The distances along paths on the surface of the TRM and angles are to be measured as the characteristics of the latent, or potential, Info-Energy of the Matrix 2-dimensional Info-Energy grid of the system. The notion of the curvature is to be defined in a way that is intrinsic to the manifold.

The Matrix 2-dimensional grid, being an external framework and boundary of the Earth enclosed at the centre of the TRM, is, simultaneously, an internal framework and

Space-Time and Info-Energy dynamic skeleton of the Earth in our dynamic world.

The 2-dimensional representations of background radiation, including those currently known as cosmic background radiation, such as the cosmic microwave background radiation, arrange the latent, or potential, Info-Energy structure of the 2-dimensional grid for the Earth and other objects and systems existing in the Universe.

The interrelationship between the 2-dimensional representations of background radiation and their dynamic representations, arranging the internal dynamic skeleton of the objects and systems in our dynamic world, provides a mechanism for the transmission of the Space-Time, Info-Energy, and Mass-Energy transformations in our dynamic world. The frequency of the electromagnetic waves and their wavelength influence the Matrix grid, changing Space-Time properties via the energy transfer.

Background radiation and light carry inertial mass from the emitting system to the absorbing system. Different forms of background radiation and light propagate with the same speed - the speed of light in vacuum. The example of the Matrix grid influence on the macro-scale is the subatomic and atomic processes associated with the regulation of heat in the body of the planet. The example of the grid-forming energy is the energy structure built by the cosmic microwave background radiation carrying this heat and information associated with the regulation of heat in the body of the Earth and other objects and systems existing in the Universe.

The potential and actual Info-Energy structures counteract and keep a dynamic balance in the Matrix, following the natural laws of thermodynamics, the principle of balance, the laws of Space-Time, Info-Energy, and Mass-

Energy Conservation, Transformations, Reversibility, Limitation, and Balance and Symmetry, or Inertia as Newton's first law and the principle of parsimony - the scientific principle that things are usually connected or behave in the simplest or most economical way.

It seems reasonable measuring the properties associated with the centre of the Matrix and characteristics of the associated system in the standard units associated with the system's volume and current understanding of time, energy, and force. Different measurement systems must be applied to the currently inaccessible Spaces of the Progressive and Regressive Time and 2-dimensional info-energetic grid forming the Matrix Spaces of the Progressive and Regressive Time.

Image 4: Earth - Moon Conjunction Image credit: NASA/JPL

Space-Time and Info-Energy Imbalance

There is no 'empty' space in the Universe, and no isolated objects and systems exist in our dynamic world. The 'outer space' is always another object or a system. Objects and systems, existing in the Universe, are tangled together by the Space-Time, Info-Energy, and Mass-Energy imbalance. The unbalanced Time-Rising radiating systems, such as high energy massive radiating systems, undergo their natural development by increasing in time. The unbalanced Space-Rising non-radiating systems, such as systems holding Black Holes and our Universe, undergo their natural development by increasing in size.

High energy massive radiating systems, such as our planet, demonstrate the Space-Time, Mass-Energy, and Info-Energy imbalance detectable in the second time dimension and the 2-dimensional time settings.

In our daily life, we apply the point-of-time settings associated with our perception of time as a point 'now' or the linear time settings following our understanding of time as linear time with the time 'flow' from the past to the future.

In the first time dimension, the object is located along the Space-Time axis at the Matrix centre associated with the period of time now occurring (Figure 1). If we apply the point-of-time and linear time dimensions to the TRM of the considered Space-Time region, there is no any sign of a gravitational field that would generate accelerated motion relative to an observer located outside of the TRM or an observer located at the Matrix centre at the '0' time point. In the first time dimension, we detect the rotation of the unbalanced multimodal systems about their Space-Time axis.

The Space-Time imbalance and the associated Info-Energy and Mass-Energy imbalance are detectable in the second time dimension (Figure 2) and 2-dimensional Time settings. In the second time dimension, we detect a new centre of the Matrix symmetry in the unbalanced, multimodal system. Two different representations of the 'o' time-point, existing in 2-dimensional time settings within the system's Matrix, indicate the Space-Time and Info-Energy imbalance.

Besides, the unbalanced Matrixes display the curved line of the structured representation of the unbalanced function that is specific to the related system. The Space-Time, Info-Energy, Arrows of Time, and Arrows of Space are not harmonically balanced in the Matrixes of these systems.

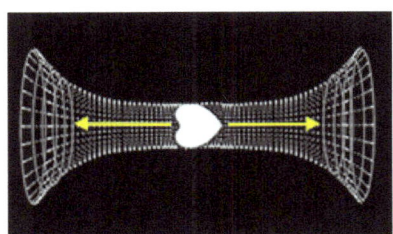

Figure 2: TRM imbalance

The excessive time, detectable in the second time dimension, is directed from the surface of high energy massive radiating systems (Figure 3). The excessive time coexists with the deficit of the system's volume. This Space-Time imbalance is supported by the Mass-Energy and Info-Energy imbalance, such as the excessive time-associated latent and potential Info-Energy of the 2-dimensional grid and the deficit of the actual Info-Energy represented in masses, kinetic energy, and other energy representations acting in the peripheral regions of the system. In modern physics, the deficit of mass and kinetic energy of the system is called 'the negative gravitational potential energy'. This Space-Time imbalance and the associated Mass-Energy and Info-Energy imbalance humans perceive as gravity on the surface of high energy massive radiating systems. For example, we can perceive gravity on the surface of the Earth and detect signs of gravity in other high energy massive radiating systems, such as stars and planets.

The deficit of time is directed from the centres of high energy massive radiating systems (Figure 4). The deficit of time coexists with the excessive, super-concentrated space of the system. This Space-Time imbalance is supported by the Mass-Energy and Info-Energy imbalance, such as the deficit of time-associated latent and potential Info-Energy of the 2-dimensional grid and the excessive actual Info-Energy represented in masses, kinetic energy, and other energy representations acting in the central region of the system in the period of time now occurring. In modern physics, the excessive mass of the system is called 'the positive mass-energy'. This Space-Time imbalance and the associated Mass-Energy and Info-Energy imbalance are reflected in antigravity.

Antigravity, prompted by the centre of the Earth, balances gravity prompted by the peripheral region of the Earth and therefore protects the planet from the gravitational shock. The projection of the different representations of the 'o' time-point, existing in the 2-dimensional time settings, into the volume of the high energy massive radiating system, such as our planet and the Sun, would draw the particular region of the Space-Time and Info-Energy imbalance with the qualities of antigravity at the centre of the system.

The Space-Time imbalance and the associated Mass-Energy and Info-Energy imbalance are reflected in the specific geometry of the high energy massive radiating objects and systems in our dynamic world. The two different representations of the '0' time-point, existing in 2-dimensional time settings, confirm the coexistence of the qualities of a toroid along with the qualities of a globe in the multimodal Space-Time and Info-Energy structures of the high energy massive radiating systems. The systems' geometry may be represented as the degeneration of a toroid in our dynamic world. We suppose that the Earth has passed the period of being non-radiating systems in the form of a toroid in the past.

Image 5: 'Earth Mars Comparison' Image credit: NASA/JPL

Matrix Forces

Every Matrix demonstrates a tendency to resist any changes. The unbalanced multimodal Space-Time and Info-Energy structures display a tendency to obtain and retain a balance and symmetry via its Space-Time axis and '0' Space-Time point at the centre of the Matrix. The Matrix imbalance resolves itself in the balancing transformations. Similarly, gravity, exposing Space-Time, Info-Energy, and Mass-Energy imbalance, results in gravitational acceleration, and antigravity is settled in anti-gravitational deceleration.

Balancing transformations are initiated by the influence of the two opposite vector-forces acting in a dynamic balance at the Matrix centre - the Matrix grid vector-force of pressure and the system's vector-force of resistance.

Balancing forces emerge within the fundamental 2-dimensional Space-Time. The Matrix grid vector-force of pressure diverges from the system's vector-force of resistance within the 2-dimensional grid arranged by background radiation. Info-Energy blocks, which are processed, stored and transmitted by the 2-dimensional grid, alter the system's structures and functioning in a manner specific to the system.

The Matrix grid vector-force of pressure applies pressure (contact force) on the surface of the system enclosed at the centre of the Matrix. This vector-force has a direction perpendicular to the surface of contact (normal force). This pressure is a measure of the system's latent, or potential, Info-Energy stored in the Matrix grid. It is related to the latent energy density and latent information density associated with the system's potential space and time of existence.

The potential space, formed within the 2-dimensional grid and containing space-associated latent Info-Energy of the system, is a transmitter between potential and actual structures of our planet. It functions similar to the restricted genetic code expression. The 2-dimensional grid keeps and transforms energy and associated information in compliance with the stored data.

The Matrix grid vector-force of pressure limits the system's volume in the Matrix Space of the Current Time and develops the system's Spaces of the Progressive and Regressive Time.

The influence of the grid vector-force of pressure upon the system's volume is associated with the Matrix tendency to transform the Space of the Current Time, or the volume of the system, into the Matrix Spaces of the Progressive and Regressive Time, along with the tendency to transform the actual Info-Energy, acting in the volume of the system as mass and kinetic energy, into the potential Info-Energy of the 2-dimensional grid, retaining the total Space-Time, Mass-Energy, and Info-Energy of the Matrix unchanged following the laws of Space-Time, Info-Energy, and Mass-Energy Conservation, Transformations, Reversibility, Limitation, and Balance and Symmetry.

The system's actual Info-Energy, bounded with the Matrix grid in the Space of the Current Time, results in the system's vector-force of resistance that influences the 2-dimensional grid (contact force). This vector-force has a direction perpendicular to the surface of contact (normal force). The system's vector-force of resistance develops the Space of the

Current Time. The vector-force of resistance is a measure of the system's actual Info-Energy acting in the volume of the system in the period of time now occurring.

The influence of the system's vector-force of resistance upon the 2-dimensional grid is associated with the Matrix tendency to transform the Spaces of the Progressive and Regressive Time into the Space of the Current Time, or the volume of the system, along with the tendency to transform the latent, or potential, Info-Energy of the 2-dimensional grid into the system's actual Info-Energy, retaining the total Space-Time, Mass-Energy, and Info-Energy of the Matrix unchanged.

The Space-Time imbalance, supported by the Mass-Energy and Info-Energy imbalance, specific geometry of the high energy massive radiating systems, and influence of the Matrix forces, acting between dissimilar dimensional layers of the system, lead to the specific rotation of these objects and systems about their Space-Time axis.

The Matrix grid vector-force of pressure and the system's vector-force of resistance initiate the rotations in opposite directions with an unequal resultant quantity and impulse of the rotation of a system. Accordingly, the resultant quantity and impulse of the rotation are unequal zero.

The specific rotation of the system about its Space-Time axis in the multimodal Space-Time is reflected in the associated rotation of the system about its axis in our dynamic world and the associated rotation of the dynamic representations of background radiation. The relative specific rotation of the Matrix grid is also reflected in the rotation of background radiation, such as cosmic background radiation.

The Matrix grid vector-force of pressure and the object's vector-force of resistance interact in a dynamic balance. Dynamics preserve the laws of Space-Time, Info-Energy, and Mass-Energy Conservation, Transformations, Reversibility, Limitation, and Balance and Symmetry, or Inertia as Newton's first law and and the principle of parsimony.

The influence of the Matrix resultant vector-force reflects the dominant influence of the Matrix grid vector-force of pressure or the object's vector-force of resistance. The influence of the Matrix vector-force at the centre of the Earth reflects the dominant influence of the Matrix grid vector-force of pressure. The influence of the Matrix vector-force in the peripheral regions of the Earth reflects the dominant influence of the system's vector-force of resistance.

Accordingly, antigravity, exposing Space-Time, Info-Energy, and Mass-Energy imbalance, results in anti-gravitational deceleration, and gravity is settled in gravitational acceleration. Dynamics preserve the natural laws mentioned above.

Image 6: 'Full Moon Over Newfoundland' Image Credit: NASA

Gravity of the Earth

The coexistence of the qualities of a toroid along with the qualities of a globe in the 2-dimensional time settings suggests the specific geometry of the planet and dynamics of the Earth historical development as the degeneration of a toroid that results in the Earth Space-Time, Mass-Energy, and Info-Energy imbalance in our dynamic world.

The specific toroid-globe geometry and the associated uneven distribution of mass are the cause of the particular geometry of a high energy massive radiating system, such as the Earth, in our dynamic world.

The Space-Time imbalance, such as the excessive time and deficit of space, accompanying Info-Energy and Mass-Energy imbalance, is mostly associated with the periphery of the high energy massive radiating system (Figure 3) and related to its geometry detectable in the second time dimension.

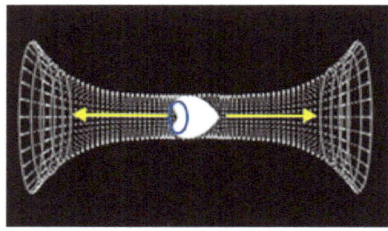

Figure 3: TRM Gravity

This specific Space-Time imbalance and the associated Mass-Energy, and Info-Energy imbalance are perceived as gravity.

We observe and experience gravity on the surface of the Earth. We can detect signs of gravity around other high energy massive radiating objects and systems, such as the Sun, other stars, planets, Globular star clusters and galaxies tangled by gravity.

The Space-Time, Info-Energy, and Mass-Energy imbalance, reflected in gravity, result in the forces providing balancing transformations reflected in gravitational acceleration.

Neither Mass-Energy nor the forces but the gravitational acceleration is the definite proof of gravity existence. The gravitational acceleration of the object's fall does not depend on the mass of the object. Mass and energy arrange the detectable physical basis for the Space-Time imbalance and balancing transformations.

The high energy massive radiating systems are not spherically symmetrical. The specific geometry of the Earth is reflected, for example, in gravitational acceleration on the surface of the Earth.

In the process of the gravitational acceleration, we can detect the unbalanced relationship between space and time as the linear space per time squared. On the Earth, objects fall with a standard value of acceleration 9.80665 m/s². Nevertheless, at different points on Earth, objects fall with an acceleration between 9.764 m/s² and 9.834 m/s² depending on altitude and latitude. The gravitational acceleration increases from about 9.780 m/s² at the equator to about 9.832 m/s² at the poles, because the relations between space and time are various at different points on the surface of the Earth and the geometry of the planet is affected by its geometry in the 2-dimensional time settings.

The gravitational acceleration, reflecting the relationship between space and time, is the element that let us recognise the gravity and the degree of gravity, measured by acceleration.

The primary process in gravitational acceleration is associated with the Space-Time transformations and the associated Info-Energy and Mass-Energy transformations.

The process, currently understood as gravity propagation, is represented in the development and concentration of the volume of the Earth and generation of mass, kinetic energy, and other energy representations filling up the volume of the Earth in the period of time now occurring.

The Matrixes of the high energy massive radiating systems, such as our planet, provide a mechanism for the Space-Time, Info-Energy, and Mass-Energy transformations and their transmission in our dynamic world.

A '0' Space-Time point of the Matrix symmetry is the point of Space-Time transformations, keeping space and time in a dynamic balance.

Two different '0' time-points coexisting in the 2-dimensional time settings, reflecting the Space-Time and Info-Energy imbalance at the centre of the Matrix, provide the basis for the Space-Time and energy transformations. The points of symmetry of the dissimilar Space-Time dimensional layers, balancing the multimodal Space-Time and Info-Energy structure of a system, are the points of the Space-Time transformations, supported by Info-Energy, and Mass-Energy transformations. The balancing quantity of flux does not emanate from the '0' space-time-energy point but diverges out of the points of symmetry of the dissimilar dimensional layers of the multimodal Space-Time and Info-Energy structure of a system.

The balancing flux of radiation diverges within the object's 2-dimensional Space-Time across the 2-dimensional Info-Energy grid.

The 'gravitational horizon' is the 'transformations horizon' - the Matrix Info-Energy grid, which is arranged by the 2-dimensional representations of background radiation. It is the leading element in the Space-Time, Info-Energy, and Mass-Energy transformations. Matrix forces arise along the 2-dimensional Info-Energy grid forming Spaces of Time. The influence of the Earth vector-force of resistance upon the Matrix Info-Energy grid develops and consolidates the volume and masses of the planet, retaining the total Mass-Energy and Info-Energy of the Matrix unchanged under the natural laws mentioned above.

The Theory of Matrix reflects Space-Time, Info-Energy and Mass-Energy transformations, associated with gravity, as follows:

A large amount of time is required in order to get a small amount of space proportional to time with the application of the Coefficient of Transformation $[c]^2$ if the Mass-Energy balance of the object is unchanged.

A large amount of energy is required in order to get a small amount of mass proportional to energy with the application of the Coefficient of Transformation $[c]^2$ if the Space-Time balance of the object is unchanged.

A large amount of latent, or potential, Info-Energy is required in order to get a small amount of actual Info-Energy proportional to latent, or potential, Info-Energy with the application of the Coefficient of Transformation $[c]^2$ if the Space-Time balance of the object is unchanged.

The Coefficient of Transformation $[c]^2$ equals the squared numerical value of the speed of electromagnetic waves propagation in a vacuum. The Coefficient of Transformation $[c]^2$ applies to the decisions associated with the Space-Time, Mass-Energy, and Info-Energy transformations, including those reflected in gravity and antigravity.

We summarise the main position related to gravity as follows. Gravity reflects the specific Space-Time imbalance, such as the excessive time and space deficit, and the associated Info-Energy and Mass-Energy imbalance, such as the excessive potential energy and the deficit of mass in the peripheral regions of the Earth and other high energy massive radiating systems.

Dynamic representations of background radiation arrange, support, transport, and transmit the balancing transformations, prompted by the peripheral regions of the Earth, associated with gravity, and reflected in gravitational acceleration, consolidation of the volume of the Earth, along with the formation of mass filling up the volume of the Earth in the period of time now occurring. Transformations of the information, depicting the forms of energy and matter, proceed along with energy transformations. Accordingly, gravity propagation reflects the process, represented in the concentration and consolidation of space and generation of masses of the Earth.

Image 7: 'NuSTAR Orbits Earth (Artist Concept)'

Image credit: NASA/JPL-Caltech

'Artist's concept showing NASA's NuSTAR mission orbiting Earth. NuSTAR will hunt for hidden black holes and other exotic cosmic objects.' NASA

Antigravity of the Earth

The massive radiating objects and systems, displaying surface gravity in our dynamic world, keep special secret balancing areas within their centres.

Antigravity keeps gravity in check. It prevents the shock-gravity collapse of our planet, the Sun, other planets, stars, and galaxies tangled by gravity. The shock-gravity collapse of a radiating system would create a 'Black hole' and the following transformation of the high energy massive radiating system into a non-radiating system.

The traditional understanding and management of gravitation did not lead us to antigravity exploration. Antigravity existence is rejected as 'impossible'. The fact of the antigravity existence must not be rejected. Anti-gravitational deceleration and associated processes must be investigated.

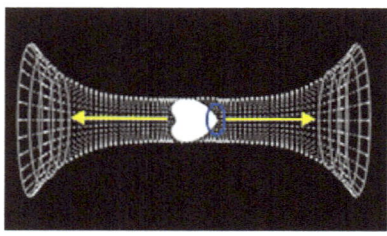

Figure 4: TRM Antigravity

Antigravity reflects the Space-Time imbalance, such as the excessive space and time deficit, and associated Info-Energy and Mass-Energy imbalance, such as the excessive mass and deficit of potential energy at the centres of the high energy massive radiating objects and systems, such as our planet.

The fundamental process, reflected in antigravity propagation within the high energy massive radiating objects, such as our planet, is associated with the transformation of the excessive, super-concentrated space at the centre of the system into the potential space and time, along with the transformation of the excessive mass and energy into the potential energy associated with the potential space and time of the system.

The generation of space, energy, and matter within the system under the influence of gravity, coexists in a dynamic balance with the degeneration of space, energy, and matter into different forms of non-radiating matter and energy prompted by antigravity.

Antigravity propagation is associated with the degeneration and reduction of space, energy, and matter prompted by the centres of high energy massive radiating systems, such as the Earth.

The observer, who could reach the centre of the Earth, would be stopped by the 'natural limit' - the impossibility of moving further to the centre and more profound.

Should we travel through the radiating Globular star cluster towards the centre (Image 8), we would experience resistance of the 'empty' space on the way as a result of space and matter deflation and concentration, along with the enforced anti-gravitational deceleration prompted by the centre of the star cluster. We could arrive at the Black Hole, with thick, relatively firm in consistency degenerated matter - 'Black Matter' around it, at the centre of the star cluster.

The Matrix of the Earth provides a mechanism for the Space-Time, Info-Energy, and Mass-Energy transformations and their transmission in our dynamic world.

The points of symmetry of the dissimilar Space-Time dimensional layers, balancing the multimodal Space-Time and Info-Energy structure of the Earth, are the points of the Space-Time transformations, supported by Info-Energy and Mass-Energy transformations.

Antigravity reflects the influence of the Matrix grid's vector-force of pressure in the central areas of the high energy massive radiating objects and systems, such as our planet.

The balancing flux of radiation diverges within the Earth 2-dimensional Space-Time across the 2-dimensional Info-Energy grid arranged by the 2-dimensional representations of background radiation. Dynamic representations of background radiation arrange the internal framework and Space-Time and Info-Energy skeleton of the Earth and other systems existing in our dynamic world.

Dynamic representations of background radiation arrange, support, transport, and transmit balancing transformations, associated with antigravity and reflected in the degeneration and reduction of space, mass, and kinetic energy reflected in the anti-gravitational deceleration prompted by the centre of the Earth in the period of time now occurring.

Space, energy, and mass, generated under the influences of the system's vector-force of resistance prompted by the peripheral regions of the high energy massive radiating system, degenerate throughout the system into different forms of non-radiating matter and energy under the influence of the 2-dimensional grid vector-force of pressure prompted by the central regions of the system, such as the Earth. The presence of non-radiating matter and energy indicates the anti-gravitational processes in the area.

We call the degenerated matter and energy, associated with anti-gravitational processes within the central regions of the high energy massive radiating systems, 'Black matter' and 'Black energy' to express the difference between the forms of degenerated matter and energy within high energy massive radiating systems and the degenerated matter and energy within non-radiating systems (known as 'Dark matter' and 'Dark energy') according to the precise direction of the time and time-associated energy flow at the centre of the Earth (Figure 4). The difference in qualities of Black matter and Dark matter is comparable to the difference in qualities of the degenerated suppressed amorphous matter with the dissociated atomic bonds at the centres of the high energy massive radiating systems and the vacuum of outer space.

Probably, the elements of the degenerated matter and energy are not so far from us as we think.

Accordingly, the process, occurring at the central areas of the Earth, can be concisely summarized as follows. Antigravity reflects the specific Space-Time imbalance, such as the excessive space and time deficit, and the associated Info-Energy and Mass-Energy imbalance, such as the excessive mass and the deficit of the potential energy at the centre of the Earth and other high energy massive radiating systems.

Antigravity propagation, arranged by the dynamic representations of background radiation, is associated with the transformation of the excessive, super-concentrated space into the potential space and time, developing inflated and undetectable potential space and time of the planet, along with the degeneration and reduction of the excessive mass in the volume of the planet in our dynamic world in the period of time now occurring.

The anti-gravitational horizon is the 'transformations horizon' - the Matrix Info-Energy grid arranged by the 2-dimensional representation of background radiation. It is the boundary of our planet in the Matrix and the dynamic internal skeleton of the Earth in our dynamic world. The dominant influence of the 2-dimensional grid vector-force of pressure provides the degeneration and reduction of space, mass, and kinetic energy reflected in the anti-gravitational processes in the area. Dynamics preserve the natural laws mentioned above.

The Theory of Matrix reflects Space-Time, Info-Energy, and Mass-Energy transformations, associated with antigravity, as follows:

A small amount of space is required in order to get a large amount of time proportional to space with the application of the Coefficient of Transformation $[c]^2$ if the Mass-Energy balance of the object is unchanged.

A small amount of mass is required in order to get a large amount of energy proportional to mass with the application of the Coefficient of Transformation $[c]^2$ if the Space-Time balance of the object is unchanged.

A small amount of actual Info-Energy is required in order to get a large amount of latent, or potential, Info-Energy proportional to actual Info-Energy with the application of the Coefficient of Transformation $[c]^2$ if the Space-Time balance of the object is unchanged.

The Coefficient of Transformation $[c]^2$ applies to the decisions associated with the Space-Time, Mass-Energy, and anti-gravitational deceleration.

The resultant vector-force of the Earth multimodal structure reflects the dominant influence of the 2-dimensional grid vector-force of pressure.

Image 8: 'Omega Centauri'

Image credit: NASA/JPL-Caltech/UCLA

The Equivalence principle

The Equivalence principle of General Relativity states that 'at any point of Space-Time the effects of a gravitational field cannot be experimentally distinguished from those due to an accelerated frame of reference' (Oxford dictionary).

We, supporting the Equivalence principle of General Relativity, view the mentioned gravitational field as the Space-Time imbalance, such as the excessive time and deficit of space supported by the associated Info-Energy and Mass-Energy imbalance, which humans perceive as gravity. At any point of Space-Time, the effects of the specific Space-Time imbalance, such as the excessive time and deficit of space, supported by the Info-Energy and Mass-Energy imbalance and reflected in gravity, cannot be experimentally distinguished from those due to an accelerated frame of reference.

We, developing the Equivalence principle of General Relativity, view antigravity as the specific Space-Time imbalance, such as the excessive space and deficit of time supported by the Info-Energy and Mass-Energy imbalance. At any point of Space-Time, the effects of the specific Space-Time imbalance, such as the excessive space and deficit of time, supported by the Info-Energy and Mass-Energy imbalance and reflected in antigravity, cannot be experimentally distinguished from those due to a decelerated frame of reference.

Transmission of Gravity and Antigravity

Any dynamics within our dynamic world is the result of the interaction of objects within a system. Being consistent in considering the gravity and antigravity propagation within our dynamic world, we refer to the law of mass interaction, such as the law of mass attraction, as a relation of objects within the multimodal system. We also refer to the law of mass interaction, such as the law of mass repulsion as the relation of objects within a system when we consider a relation of objects which tend to move away from each other in our dynamic world.

Accordingly, the forces acting in the multimodal Space-Time and Info-Energy structure of the Earth are attractive and repulsive, similar to electrostatic forces, acting under an inverse-square law, analogous to Isaac Newton's inverse-square law of universal gravitation. These forces, attractive and repulsive, provide the physical basis for gravity and antigravity propagation within the high energy massive radiating system, such as the Earth, existing in our dynamic world.

Gravity propagation is associated with the propagation of the generation and concentration of space and mass, prompted by the Space-Time, Mass-Energy, and Info-Energy imbalance in the peripheral areas of the Earth.

Antigravity propagation is associated with the propagation of the degeneration and reduction of space and mass prompted by the Space-Time, Mass-Energy, and Info-Energy imbalance at the centre of the Earth.

The gravitational and anti-gravitational horizon is the Matrix Info-Energy grid, which is the boundary of our planet in the multimodal Space-Time and the dynamic internal Info-Energy framework of the Earth in our dynamic world.

The Earth 2-dimensional grid represents the 2-dimensional Space-Time structure of the latent Info-Energy of our planet arranged by the 2-dimensional representation of background radiation.

The Earth 2-dimensional grid act as a transmitter, similar to a membrane of a living cell and structural genes, between time-associated latent Info-Energy and space-associated actual Info-Energy of the planet. It keeps and transforms energy and associated information in compliance with a stored data to balance an existing capacity to perform work against its actual realisation in forms of actual Info-Energy, such as mass, kinetic energy, and other energy representations, acting in the volume of the Earth in the period of time now occurring.

A capacity to perform work is being realized under the influence of the imbalance between actual Info-Energy and latent Info-Energy on the one hand. On the other hand, the imbalance between Info-Energy of potential space and time-associated Info-Energy influences the planetary dynamics.

The potential and actual Info-Energy structures counteract and keep a balance in the Matrix, retaining the Laws of Space-Time, Mass-Energy and Info-Energy Reversibility, Limitation, Transformation, Conservation, and Symmetry.

Although, radiation within 2-dimensional Space-Time of an object is neither progression nor emission but a divergence activating two opposite vector-forces acting in a dynamic balance and directing the balancing transformations.

'Relativity theory ... shares with the corpuscular theory of light the unusual property that light carries inertial mass from the emitting to the absorbing object.' [Einstein Albert, The Development of Our Views on the Composition and Essence of Radiation (1909)].

The 2-dimensional representations of background radiation, arranging the 2-dimensional grid, interrelate with dynamic representations of background radiation. This interrelationship provides a mechanism for the transmission of Space-Time, Info-Energy, and Mass-Energy transformations in our dynamic world. Background radiation and light carry inertial mass from the emitting to the absorbing object.

Space-Time, Info-Energy and Mass-Energy transformations propagate with the speed of light and are carried by background radiation. The frequency of the electromagnetic waves and their wavelength influence the Earth 2-dimensional Info-Energy grid, changing Space-Time, Mass-Energy, and info-Energy properties of the Earth via the energy transfer.

The speed of gravity propagation was introduced by Hendrik Lorentz (1900) and confirmed by Hermann Minkowski in his work 'The Fundamental Equations for Electromagnetic Processes in Moving Bodies, Appendix' (1908): 'The law of mass attraction, which has been just described and which is formulated in accordance with the relativity postulate, would signify that gravitation is propagated with the velocity of light.'

We develop the theory of light and introduce the Space-Time Coefficient [c]. The Space-Time Coefficient [c] equals the numerical value of the velocity of light and electromagnetic waves propagation in a vacuum. The value of the speed of light in a vacuum (c) is 299 792 458 m s-1 (meter per second). It is a CODATA Fundamental Physical Constant.

The speed of light in vacuum sets the fundamental relation between 1-dimensional space and 1-dimensional time that defines the exact proportion of space to time as the upper limit of our dynamic world.

The 1-dimensional Space-Time, supported by energy, is an undifferentiated existence of space, time, and energy underlying our dynamic world. It is built by and filled with the one quality Info-Energy of infinite duration, presenting our dynamic world with the ultimate dynamics in accordance with the human perception of the relation of 1-dimensional space to 1-dimensional time as speed. Speed as the rapidity of movement represents the relation between 1-dimensional space and 1-dimensional time.

The Space-Time Coefficient is a factor that measures the exact proportion of this relation. This fundamental relation is supported by the energy divergence between space-associated actual energy and time-associated potential energy.

Planck length, Planck time, and Planck energy, based on the calculations using the speed of light in vacuum as the fundamentals proven by experiments, reflect the minimal Space-Time and Info-Energy imbalance in the existing objects and systems as the necessary deviations from the '0' Space-Time point and '0' energy point. These Planck units define the exact lower limits as the conditions of an object or a system's existence and, accordingly, factors supporting the reverse of the Space-Time and Info-Energy flow in the thermodynamic Universe and the existing objects and systems as we measure and sense them. If a system has reached the limits actuating the reverse, the flow of Space-Time and Info-Energy within the system is being reversed by the Matrix forces.

The concept of relativity and Einstein's Special Theory of Relativity state that all motion is relative and that the velocity of light in a vacuum has a constant value which nothing can exceed.

We add that the numerical value of the velocity of light in a vacuum is the Space-Time Coefficient that defines the relation between 1-dimensional space and 1-dimensional time in a vacuum. The speed of light defines the dynamic body of the multimodal Universe in our dynamic world.

The Space-Time Coefficient [c] applies to the decisions associated with the transmission of the Space-Time, Mass-Energy, and Info-Energy transformations, including those reflected in gravity and antigravity propagation in our dynamic world.

Accordingly, balancing transformations, caused by the influence of the Matrix forces and reflected in the reorganisation of space, time, matter, and energy, resulting in the generation and concentration of volumes and mass prompted by the peripheral regions of our planet, along with the degeneration and reduction of volumes and mass prompted by the centre of our planet, propagate with the speed of light in our dynamic world.

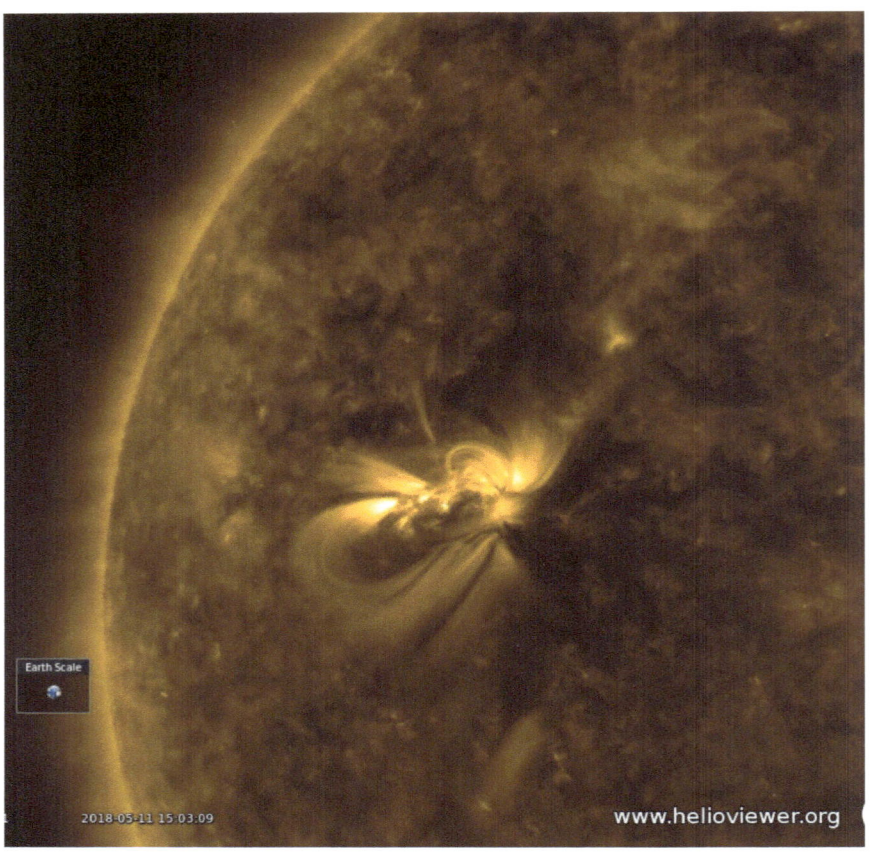

Image 9: 'Coronal Hole Facing Earth' Image credit: NASA/GSFC/Solar Dynamics Observatory

'An extensive equatorial coronal hole has rotated so that it is now facing Earth (May 2-4, 2018). The dark coronal hole extends about halfway across the solar disk. It was observed in a wavelength of extreme ultraviolet light. This magnetically open area is streaming solar wind (i.e., a stream of charged particles released from the sun) into space. When Earth enters a solar wind stream and the stream interacts with our magnetosphere, we often experience nice displays of aurora.' NASA

Rotation of the Earth

On the one hand, the object can spin without gravity. On the other hand, the object displays gravity or antigravity only if it does spin on its axis, or say, no spinning - no gravity, neither antigravity.

Would you argue that the Moon does rotate?

'The Moon does spin on its axis, completing a rotation once every 27.3 days; the confusion is caused because it also takes the same period to orbit the Earth, so that it keeps the same side facing us.' NASA

The Moon both orbits counterclockwise around Earth and rotates counterclockwise on its axis. It also takes approximately 27 days for the Moon to rotate once on its axis and 27.3 days to orbit the Earth. As a result of the synchronous rotation, the Moon appears to observers from Earth to be keeping still.

'The Moon has no side that is constantly dark; the front and back are alternately lit as the Moon rotates. Far side is a more accurate term.' NASA

We can compare the rotation of the Moon with the rotation of the Earth and other planets of the Solar system.

Earth spins on its North Pole-South Pole axis once every 24-hour day. It rotates counterclockwise on its axis with roughly the 1000 mile per hour rotation speed at the equator.

The planets all revolve around the Sun in the same direction. They orbit the Sun in a counterclockwise direction as viewed from above the Earth Geographic North Pole where Earth's axis of rotation meets its surface in the Northern Hemisphere.

Most planets, with the exceptions of Venus and Uranus, also rotate on their axes in a counterclockwise direction.

Venus rotates clockwise in retrograde rotation. What else is different in Venus rotation? Venus rotates once every 243 Earth days. It is the slowest rotation of any planet. Besides, Venus is very close to the spherical form.

We remember that the Space-Time and Info-Energy imbalance that we perceive as gravity is reflected in the geometry of the system, such as Venus, and we can see that the difference between the '0' time-points in the Venus 2-dimensional time settings is much smaller than in the Earth 2-dimensional time. Accordingly, the Space-Time and Info-Energy imbalance in Venus is much smaller than on the Earth. Gravity on the surface of Venus must be smaller than gravity on the surface of the Earth. Besides, the layer of Space-Time, affected by gravity, must be narrower than the region affected by gravity on the Earth.

The spherical form of the planet and its slow rotation let us think about the lower Space-Time and Info-Energy imbalance and lower value of gravitational acceleration on the surface of the planet.

The specific toroid-globe geometry of the system's multimodal structure is reflected in the uneven distribution of mass and the particular geometry of the system, such as the Earth, in our dynamic world.

The system's Space-Time imbalance and associated Info-Energy and Mass-Energy imbalance activate the two opposite contact vector-forces - the Info-Energy grid vector-force of pressure and the system's vector-force of resistance, which are responsible for the rotation of the unbalanced system, such as planets, about the axis of rotation.

Please consider an example. Schrödinger equation for a particle, encountering a rectangular potential energy barrier, describes the conditions for a multimodal particle at the centre of its Matrix, at the '0' time-point - the point 'now'. Under the Theory of Matrix, a multimodal particle, which impinges on the barrier from one side, demonstrates the Space-Time and Info-Energy imbalance, similar to other high energy radiating objects and systems, such as planets and stars. The Space-Time imbalance and the associated Info-Energy imbalance are reflected in gravity or antigravity - in case of a microscopic hole.

The rectangular barrier is associated with the 2-dimensional rectangular Info-Energy grid arranged by the 2-dimensional representation of electromagnetic radiation. The change of the frequency of the electromagnetic waves would be an elegant, practical, real-life decision of the problems associated with the rectangular barrier. This influence on the Space-Time and Info-Energy imbalance of the blocked particle will change the properties of its 2-dimensional rectangular Info-Energy grid, and accordingly, change the particle behaviour, including such qualities as rotation and gravity.

The influence of the vector-forces provides the balancing Space-Time, Info-Energy and Mass-Energy transformations. Balancing transformations, reflected in the reorganization of space, time, matter, and energy, have a secondary effect on the rotation of the unbalanced radiating and non-radiating objects and systems in Macro-world.

The Matrix grid vector-force of pressure, prompted by the centre of the Earth, and the system's vector-force of resistance, prompted by the peripheral regions of the Earth, initiate the rotations about the system's Space-Time axis in the opposite directions with an unequal resultant quantity and impulse of the rotation of the system in the multimodal Space-Time. Accordingly, the resultant quantity and impulse of the rotation are unequal to zero.

The pressure of the Matrix forces affects and direct the associated rotation of the 2-dimensional grid. The specific rotation of the 2-dimensional grid, being at the same time the external framework of the Earth in the multimodal Space-Time and the internal dynamic skeleton of the Earth in our dynamic world, is reflected in the rotation of background radiation, producing the rotating field of the Earth with the moving polarities in which its opposite poles rotate about a central point projected in the volume of the planet that must be detectable. The central point, projected in the volume of the planet, reflects the 'o' time-point existing in the TRM 2-dimensional time settings in the multimodal Space-Time and Info-Energy structure of the Earth.

If to translate the process, described above, into the conventional language, the process to compare with would be the following: the rotation of electromagnetic radiation, producing the rotating magnetic field of the Earth, activates a dynamo, converting mechanical energy into electrical energy at the centre of the Earth. Simultaneously, the electromagnetic radiation, rotating in the opposite direction and producing the rotating magnetic field of the Earth, activates an electric motor dominating all over the surface of the Earth and converting electrical energy into mechanical energy. In case of gravity and antigravity of the Earth and other high energy massive radiating systems, the Matrix of the system is not the field but the multimodal Space-Time and Info-Energy structure. The process is located within the curvature of Space-Time affected by the Earth gravity and the uneven distribution of mass.

The Earth spins, and its rotation is determined by the Earth resultant vector-force - the 2-dimensional grid vector-force of pressure dominating at the centre of the planet. Its influence, projected on the surface of the Earth, is the strongest at the equator, partially neutralising a degree of gravity. Accordingly, the gravitational acceleration decreases from about 9.832 m/s² at the poles to about 9.780 m/s² at the equator.

You can argue that the gravitational acceleration decreases at the equator as a result of the influence of the centrifugal force in the rotating frame of reference or because the points on the equator are furthest from the centre of the Earth.

Shall we look at the Sun?

The Sun (Image 9) spins on its axis and revolves around the centre of the Milky Way galaxy. The Sun is 'a hot ball of glowing gases at the heart of our solar system', and it also spins faster at its equator than at the poles.

'Since the Sun is a ball of gas/plasma, it does not have to rotate rigidly like the solid planets and moons do. In fact, the Sun's equatorial regions rotate faster (taking only about 24 days) than the polar regions (which rotate once in more than 30 days). The source of this "differential rotation" is an area of current research in solar astronomy.' NASA

Furthermore, the inner parts of the Sun spin faster than the outer layers. Isn't it exciting?

The inner parts of the Sun, located closer to the centre, spin faster than the outer layers, which are located further away from the centre of the Sun and being influenced by the centrifugal force. Accordingly, the centrifugal force is not the only game in town but the secondary influence on the rotation of the Sun and the planets.

The Earth rotates, and its rotation is determined by the Space-Time and Info-Energy imbalance, reflected in gravity and antigravity, and balancing transformations, reflected in gravitational acceleration and anti-gravitational deceleration, which are initiated by the two balancing forces, acting between dissimilar dimensional layers of the Earth multimodal structure.

The balancing forces are directly responsible for the rotation of the system.

The relative specific rotation of the 2-dimensional grid, being the external framework of the Earth in the multimodal Space-Time and its internal dynamic framework in our dynamic world, is reflected in the rotation of background radiation.

The specific rotation of the system about its Space-Time axis impacts the associated rotation about the axis of rotation in our dynamic world with the gradual shift in the orientation of the axis of rotation, or the precession.

Space sometimes is looking flat and time moves the world. Masses, associated with the visible space, are rotated by the centrifugal force inside the temporal Matrixes, around their mass centres, at '0' time-points of our understanding, like the surface of water rotating in a bucket. Space-Time flux changes properties and builds channels, path shorting and bringing forth spaces close together or further apart.

We mentioned above that the unbalanced system might develop the transmitting Black Holes. The transmitting Black Hole is a contact path equilibrating the system.

The central Black Hole connects the central points of symmetry of the associated Matrixes and the shared 2-dimensional Space-Time areas of the Matrixes connection.

The contact path, equilibrating the high energy massive radiating system, such as the Sun or our planet, might be developed by the surface of the system, if the anti-gravitational processes associated with the degeneration and reduction of space and mass, prompted by the centre of the system, would dominate the gravitational processes prompted by the system's peripheral regions.

The surface orientated Black Holes would be transmitting volumes, matter, and energy to the surface of the unbalanced system holding these Black Holes. The transformations within the transmitting Black Holes would demonstrate that the latent, or potential, Info-Energy of the Matrix 2-dimensional grid is being processed and transformed by the Black Hole into kinetic energy and mass of the particular Space-Time region, which this Black Hole develops.

Alternatively, the surface orientated Black Holes release energy and matter under the influence of the anti-gravitational processes prompted by the centre of the system. The 2-dimensional grid vector-force of pressure, generating anti-gravitational processes at the centre of the system, can project its influence via dynamic representations of background radiation on to the surface of the system with possible creation of the radiating Black Holes on the surface of the system. Its influence, projected on the surface of the system, is the strongest at the equator, partially neutralising a degree of gravity. We can expect these radiating Black Holes to be formed in the equatorial regions of the system. (Image 10).

The rotation of the unbalanced system impacts the rotation of the associated Black Hole, and the flow of matter, volumes, and energy within the Black Hole.

A Black Hole evaporates if the system is balanced in its Space-Time, Info-Energy, and Mass-Energy and the resultant Matrix force equals zero.

If the system has reached the lower limits for an object or a system's existence, such as Planck length, Planck time, and Planck energy, it will be reversed. The reverse of an unbalanced multimodal system affects the specific rotation of this system and the associated rotation of the background radiation. The newly established Space-Time and Info-Energy imbalance can change the direction of rotation about the system's Space-Time axis to the opposite and impact the rotation of the reversed system in our dynamic world. The rotation of the reversed system about its axis of rotation would reflect a new direction of its Space-Time axis and, accordingly, a new axis of rotation in our dynamic world. It would impact on the gradual shift in the orientation of the system's axis of rotation. The reverse of an unbalanced multimodal system, holding the Black Hole, affects the rotation of the Black Hole and the flow of matter, volumes, and energy, which this Black Hole transmits, along with the rotation of the associated Space-Time region.

Afterword

The Theory of Matrix is a new theory combining elements of psychology, cosmology, and astrophysics. This theory was introduced by Dr Audrey E. Randles in her work 'The Theory of Matrix' in 2012.

Dr Randles developed the program for psychological investigations of the Universal Matrix along with the development of the Coresynthesis Psychological Model in early 1990s. There was a space of ten years between the first explanations of the results and association of these results with the characteristics mathematically introduced by Lorenz, Minkowski, and Einstein for the Light Cone.

When Dr Randles published her work, the Light Cone was considered a specific case that can be applied solely to the flash of light. She brought forward the idea of the universality of the Space-Time structure of the objects and a new vision on Space-Time physics.

Following the analysis of the parallels and variations between the Universal Matrix and Light Cone Dr Randles calls the Light Cone 'the Matrix of Light' and presents the Theory of Matrix as the relative importance for the true understanding of the multimodal world.

The Theory of Matrix introduces the new understanding of the multidimensional world with respect to multidimensional time. Time is no longer seen as another dimension of space, nor as a momentary feature of an event but as a multidimensional element in its own right. Dr Randles associates space with Actuality and time with

Potentiality and Latency. Therefore, the objects are viewed as multimodal, multidimensional objects in Space-Time.

Books on Cosmology by Audrey E. Randles:
'Systems Theory in Cosmology' (2020)
'The Multimodal World' (2020)
'Black Holes and Supernovas' (2016)
'Grand Universe' (2016)
'Antigravity' (2015)
'Supernovas' (2015)
'The Primary Black Hole of the Universe' (2015)
'Energy in Cosmology' (2014)
'Gravity and Antigravity to the Point' (2014)
The Theory of Matrix series of books (2012 - 2013) includes the following books:
'Blocks of the Universe'
'Space and Time'
'Energy of Existence'
'Gravity and Rotation'
'Black Holes'
'Matrix of the Universe'
The New 2020 books on Cosmology include Kindle ebook and a paperback of the same title at Amazon's Book Store.

Content Use Policy

© Audrey Elizabeth Randles

Content may be used for any purpose without prior permission, subject to the special cases noted below.

By downloading the material, the user agrees:

1. to use a credit line in connection with the content. Unless otherwise noted in the caption information for any content and images the credit line should be

"Audrey E. Randles, 'The Theory of Matrix. Gravity and Rotation' (2013) red. 2020".

2. that we do not represent others who may claim to be authors or owners of copyright of any of the content, and make no warranties as to the quality of the content;

3. that we shall not be responsible for any loss or expenses resulting from the use of the content, and you release and hold us harmless from all liability arising from such use.

Special Cases:

This content is available for educational, journalistic, personal uses and scientific research following a scientific code of ethics.

Restrictions are placed on commercial uses. To obtain permission for commercial use, contact the copyright owner Dr Audrey E. Randles.

www.ingramcontent.com/pod-product-compliance
Lightning Source LLC
Chambersburg PA
CBHW040237220526
45473CB00001B/279